NIST HANDBOOK 153
2009 Edition

# LABORATORY RECOGNITION PROCESS FOR PROJECT 25 COMPLIANCE ASSESSMENT

Kurt B. Fischer and Andrew Thiessen, Editors

Office of Law Enforcement Standards
Electronics and Electrical Engineering Laboratory

January 2009

U.S. Department of Commerce
Carlos M. Gutierrez, Secretary
National Institute of Standards and Technology
Patrick D. Gallagher, Acting Director

## CONTENTS

# FOREWORD

Project 25 (P25) is a standards development process for the design, manufacture and evaluation of interoperable digital two-way wireless communications products created for public safety professionals. The published P25 standards suite is administered by the Telecommunications Industry Association (TIA Mobile and Personal Private Radio Standards Committee TR-8). Radio equipment that demonstrates compliance with P25 should meet a set of minimum requirements to fit certain needs of public safety, including interoperability, allowing users on different systems to talk via direct radio contact.

The P25 Compliance Assessment Program (CAP) was established on the basis of requests from the United States Congress[1] and is a voluntary system that provides a mechanism for the recognition of testing laboratories based on internationally accepted standards. This handbook sets forth the procedures and general requirements under which the P25 CAP operates as a third party to recognize P25 equipment testing laboratories. Although voluntary, the actions of the P25 CAP and its related laboratory recognition process are emphasized by the use of the infinitive "will" and the actions of the laboratory by the use of the infinitive "shall" to show that these actions are obligatory and requirements of the program.

Access to P25 CAP laboratory recognition is available to public and private testing laboratories, including commercial laboratories, equipment manufacturers' in-house laboratories, university laboratories, and federal, state, and local government laboratories from inside or outside the United States.

This handbook is sponsored by the NIST Office of Law Enforcement Standards' (OLES') Public Safety Communications Systems program, which provides technical expertise to the DHS Office of Interoperability and Compatibility that is building a framework for fully interoperable communications among all first response agencies. It is also developing standards for voice, data and image transfers, evaluating existing devices and services, and developing a uniform assessment process for testing the conformity of mobile radio equipment with P25 requirements.

This laboratory recognition process is not part of the NIST National Voluntary Laboratory Accreditation Program (NVLAP). However, the editors thank NVLAP for providing a model for the assessment component of the P25 CAP Laboratory Recognition Process.

The revisions made for this edition of Handbook 153 were in two primary categories:

1. Per the DHS P25 CAP Charter, the roles of the CAP Laboratory Program Manager versus the OIC CAP/Program Manager are clarified.

---

[1] Senate Report 109-088 - DEPARTMENTS OF COMMERCE AND JUSTICE, SCIENCE, AND RELATED AGENCIES APPROPRIATIONS BILL, 2006 and House Report 109-241 - MAKING APPROPRIATIONS FOR THE DEPARTMENT OF HOMELAND SECURITY FOR THE FISCAL YEAR ENDING SEPTEMBER 30, 2006, AND FOR OTHER PURPOSES.

2. The original NIST Handbook 153:2007 referenced Technical Service Bulletins as the key documents that will guide testing procedures in the P25 CAP. This version revises that to point readers to the DHS Compliance Assessment Bulletins (CABs), per the DHS P25 CAP Charter, which themselves leverage information from existing P25 Technical Service Bulletins (TSBs).

No revisions have been made to this edition that would alter the technical requirements on laboratories to become recognized under the P25 CAP.

This publication supersedes NIST Handbook 153:2007.

# 1. GENERAL INFORMATION

## 1.1 Normative References

The following references are required for the application of this handbook. For dated references, only the edition cited applies. For undated references, the latest edition of the referenced document applies.

> Title 15 United States Code, Section 3710a. Cooperative research and development agreements.
>
> ISO/IEC 17050:2004, Conformity assessment — Supplier's declaration of conformity.
>
> P25 Compliance Assessment Bulletin (CAB) — Issued by US Department of Homeland Security (DHS), Office for Interoperability and Compatibility (OIC).

## 1.2 Informative References

The following references are important for the application of this handbook. For dated references, only the edition cited applies. For undated references, the latest edition of the referenced document applies. The last two references are to publications not yet available at the time of the publication of this handbook.

> ISO/IEC 17000:2004, Conformity assessment — Vocabulary and general principles
>
> ISO/IEC 17011:2004, Conformity assessment — General requirements for accreditation bodies accrediting conformity assessment bodies
>
> ISO/IEC 17025:2005, General requirements for the competence of testing and calibration laboratories
>
> NIST Handbook 150:2006, National Voluntary Laboratory Accreditation Program Procedures and General Requirements
>
> NIST Document, Project 25 Compliance Assessment Program Test Method Review Summary
>
> Compliance Related Telecommunications Systems Bulletins

## 1.3 Terms and Definitions

**1.3.1** *American Association for Laboratory Accreditation (A2LA)*
A2LA provides comprehensive services in accreditation and training for testing and calibration laboratories. See also NVLAP below.

**1.3.2** *Asia Pacific Laboratory Accreditation Cooperation (APLAC)*

APLAC is a group of accreditation bodies in the Asia Pacific region responsible for accrediting calibration, testing and inspection facilities.

**1.3.3** *Authorized Representative*

An individual who is authorized to commit a Laboratory to fulfill the P25 Compliance Assessment Program's conditions for recognition.

**1.3.4** *Certificate of Recognition*

A document presented by the P25 CAP to a laboratory that has demonstrated competence to conduct a particular Scope of Recognition of testing, that is, all of the test cases or a subset of test cases defined in the applicable Compliance Assessment Bulletin.

**1.3.5** *Compliance Assessment Bulletin*

A Compliance Assessment Bulletin contains the policies and procedures by which the P25 CAP operates. A Compliance Assessment Bulletin is approved by the Project 25 Compliance Assessment Program Governing Board for publication by DHS OIC.

**1.3.6** *Compliance Related Telecommunications Systems Bulletin*

A Telecommunications Systems Bulletin (see below) published by the Telecommunications Industry Association (TIA) that defines a particular set of test methods required to demonstrate that equipment built to a particular interface standard is compliant with those standards.

**1.3.7** *Equipment Supplier*

The Original Equipment Manufacturer (OEM) or an authorized agent of the OEM.

**1.3.8** *International Laboratory Accreditation Cooperation (ILAC)*

An international cooperation of laboratory and inspection accreditation bodies that provides a focus for:

a) Developing and harmonizing laboratory and inspection accreditation practices

b) Promoting laboratory and inspection accreditation to industry, governments, regulators and consumers

c) Assisting and supporting developing accreditation systems

It also provides global recognition of laboratories and inspection facilities via the ILAC Mutual Recognition Arrangement (see http://www.ilac.org/).

**1.3.9** *Invitational Testing Session*

A group testing session hosted by an infrastructure equipment provider for the purposes of evaluating equipment interoperability.

**1.3.10** *Laboratory Assessment Team*

One or more individuals possessing knowledge of ISO/IEC 17025 processes and technical familiarity with a particular Scope of Recognition of testing who travel to a laboratory to assess its competence to prescribed standards. Although the term Team is used here, a lead assessor may possess the competence to also function as the subject matter expert and thus act as a "single person" Laboratory Assessment Team.

**1.3.11** *Laboratory Client*

The entity that contracts with the laboratory; usually the laboratory client is the Equipment Supplier.

**1.3.12** *Laboratory Code*

A unique identifier for program management assigned to each laboratory participating in the P25 Compliance Assessment Program.

**1.3.13** *Model Class*

Products defined by the manufacturer as having identical P25 functionality shall constitute a Model Class.

**1.3.14** *Monitoring Visit*

An on-site assessment of a laboratory not associated with an initial or renewal assessment.

**1.3.15** *Mutual Recognition Arrangement (MRA)*

A signed document concluded among governmental and other authorities of different countries for mutual recognition of accreditations.

**1.3.16** *National Voluntary Laboratory Accreditation Program (NVLAP)*

NIST NVLAP provides third-party accreditation to testing and calibration laboratories. NVLAP's accreditation programs are established in response to Congressional mandates or administrative actions by the Federal Government or from requests by private-sector organizations. See A2LA in Section 1.3.1.

**1.3.17** *P25 Compliance Assessment Program/Laboratory Program Manager (P25 CAP/LPM)*

The Program Manager for the laboratory assessment and provides a recommendation to DHS for the recognition portion of the P25 Compliance Assessment Program.

**1.3.18** *DHS Office of Interoperability and Compatibility (OIC) Compliance Assessment Program Manager (OIC CAP/PM)*

The Program Manager for the Laboratory Recognition portion of the P25 CAP.

**1.3.19** *P25 Compliance Test Method*

A normative test procedure defined for compliance testing of a given interface. These test procedures are typically broken down into the following categories:

a) Performance (e.g., Measurement Methods and Performance Recommendations)

b) Conformance

c) Interoperability

**1.3.20** *Scope of Recognition*

The type of testing for which a laboratory has demonstrated compliance and competence. The particular test methods are contained in the relevant Compliance Assessment Bulletin approved by the P25 CAP Governing Board and published by DHS. The Scope of Recognition will list the actual test methods for which the laboratory has demonstrated competency. NOTE: The Scope of Recognition need not include all of the test methods contained in the aforementioned CABs.

**1.3.21** *Supplier's Declaration of Compliance (SDoC)*

A formal declaration of compliance created in accordance with ISO/IEC 17050 for a particular set of P25 Compliance Test Methods defined within the relevant Compliance Assessment Bulletin(s). The SDoC is signed by an Authorized Representative of the Equipment Supplier. More information on the SDoC is contained in DHS SDoC CAB.

**1.3.22** *Telecommunications Systems Bulletin (TSB)*

Telecommunications Systems Bulletins are published by the Telecommunications Industry Association (TIA). A TSB is not a standard, but rather contains technical material that may be valuable to industry and users.

**1.3.23** *Test Case*

A particular section that defines a unique test procedure within a compliance test methods document.

**1.3.24** *Test Method Review Summary*

A document template used by members of a P25 CAP Laboratory Assessment Team to record their observations of the management system and demonstrations of test procedures during an on-site assessment.

**1.3.25** *Test Report Format*

A predefined format for summary test reports required by the P25 CAP for use by Equipment Suppliers participating in the recognition program.

## 1.4 Purpose and Scope

**1.4.1** This handbook sets forth the procedures and general requirements under which the Project 25 Compliance Assessment Program operates as an unbiased third party to recognize Project 25 equipment testing laboratories.

**1.4.2** The required P25 Compliance Test Methods and Test Cases are defined in TIA standards and Compliance Assessment Bulletins.

**1.4.3** This handbook constitutes the collective body of requirements that must be met by a laboratory seeking P25 CAP recognition for any of the specific test methods (for example, audio level) provided in the laboratory's Scope of Recognition. A given laboratory's Scope of Recognition may or may not cover all of the TIA-defined needs or recommendations.

### 1.5 Outline of the P25 Compliance Assessment Program

**1.5.1** The P25 Compliance Assessment Program (CAP) is a voluntary system that provides a mechanism for the recognition of testing laboratories based on internationally accepted standards. It identifies competent laboratories through assessments by trained Laboratory Assessment Teams (see Section 2.2) and promotes the acceptance of compliant test results from these laboratories.

**1.5.2** The P25 CAP was established on the basis of requests from the United States Congress[2]. The specific tests, types of tests, and standards to be included in the program are defined in Compliance Assessment Bulletins approved by the Project 25 CAP Governing Board. The laboratory assessment part of the program is managed by the Project 25 Compliance Assessment Program Laboratory Program Manager (P25 CAP/LPM) (see Section 2.1.1) who reports periodically to and receives feedback from government stakeholders. The recognition component of the program is managed by the DHS OIC CAP Program Manager who reports and receives feedback from Government stakeholders. While the P25 CAP/LPM and the OIC CAP/PM exercise a great deal of authority and autonomy over the program, these individuals do not unilaterally propose or decide the scope of the program.

**1.5.3** The P25 CAP/LPM administers laboratory related policies and procedures in a non-discriminatory manner. Access to P25 CAP laboratory recognition is not conditioned on the size of a laboratory or on its membership in any association or group, nor is it conditioned upon the number of laboratories already recognized. P25 CAP services are available to public and private testing laboratories, including commercial laboratories, manufacturers' in-house laboratories, university laboratories, and federal, state, and local government laboratories. For laboratories operating outside the United States, the P25 CAP/LPM may accept reports prepared by Laboratory Assessment Teams operating under ISO/IEC 17011 accreditation bodies that have signed a Mutual Recognition Arrangement with APLAC and/or ILAC. The provisions of this Handbook shall govern the conduct of all laboratory assessments irrespective of the composition of Laboratory Assessment Teams (see Section 1.4.3).

**1.5.4** P25 Compliance Assessment Program laboratory recognition is based on evaluation of a laboratory's quality management system and technical competence for conducting specific test methods and measurements in certain fields or scopes of testing. Recognition is granted only after an applicant has demonstrated that it has met all P25 CAP laboratory requirements in this handbook. Recognition is acknowledged by the

---

[2] Senate Report 109-088 - DEPARTMENTS OF COMMERCE AND JUSTICE, SCIENCE, AND RELATED AGENCIES APPROPRIATIONS BILL, 2006 and House Report 109-241 - MAKING APPROPRIATIONS FOR THE DEPARTMENT OF HOMELAND SECURITY FOR THE FISCAL YEAR ENDING SEPTEMBER 30, 2006, AND FOR OTHER PURPOSES.

issuance of a Certificate and Scope of Recognition, which details the specific test methods, measurements and services for which a laboratory has been recognized. DHS recognizes the laboratory based on a nomination from the P25 CAP LPM. DHS is free to independently assess the application package from the P25 CAP LPM and is not bound by the LPM's recommendation.

**1.5.5** The P25 CAP/LPM operates a management system that is compliant with ISO/IEC 17011:2004; however, this system does not undergo peer review and is not part of the APLAC and/or ILAC Mutual Recognition Arrangements (MRAs).

**1.5.6** P25 CAP laboratory recognition does not relieve a laboratory from complying with applicable federal, state, and local laws and regulations.

## 1.6 Confidentiality

**1.6.1** To the extent permitted by applicable laws, the P25 CAP will protect the confidentiality of all information obtained relating to the application, on-site assessment, evaluation, and recognition of laboratories.

**1.6.2** In addition, P25 CAP and the laboratory further agree that, to the extent permitted by law, P25 CAP/LPM and Laboratory Assessment Team members will protect information obtained during application, on-site assessment, evaluation, and recognition from disclosure pursuant to Title 15 USC 3710a(c)(7)(A) and (7)(B) for a period of five years after such information is obtained.

**1.6.3** For the first five years that laboratory information is held by the P25 CAP, the provisions of sections 1.6.1 and 1.6.2 will be in force. Information in the P25 CAP's possession for more than five years will continue to be held in confidence under the provisions of Section 1.6.1.

## 1.7 Complaints

The P25 CAP/LPM will document the system used to address complaints, which will include the procedures for determining the validity of complaints, taking appropriate and effective actions, responding to complainants, and record-keeping. A complaint regarding the activities of the P25 CAP or of a P25 CAP-recognized laboratory may be lodged by any person or organization. Information about the complaint shall be put in writing and mailed, faxed, or e-mailed to the P25 CAP/LPM if related to laboratory assessments or to the OIC CAP/PM if related to laboratory recognition status, along with supporting documentation, if available. A complaint concerning a P25 CAP-recognized laboratory shall first be submitted to the laboratory against which the complaint is lodged. The P25 CAP LPM will notify, if appropriate, the OIC CAP/PM unless the complainant requests otherwise.

# 2. ORGANIZATION

## 2.1 Management of the Project 25 Compliance Assessment Program

**2.1.1** P25 Compliance Assessment Program Laboratory Program Manager (P25 CAP/LPM)

**2.1.1.1** *Responsibilities*

The P25 CAP/LPM oversees the day-to-day operations of the program and is responsible for:

a) Selecting members of Laboratory Assessment Teams

b) Recommending to OIC CAP/PM qualified laboratories as candidates for recognition

c) Facilitating the resolution of laboratory-related or testing-related disputes between parties to the P25 CAP

d) Briefing government stakeholders on a periodic basis

e) Adapting the management of the program to address the changing needs of P25 users as communicated by the local, state, and Federal government user and stakeholder representatives on the DHS P25 CAP Governing Board.

**2.1.1.2** *Selection*

The P25 CAP/LPM is selected by the Director of NIST Office of Law Enforcement Standards (OLES). A deputy P25 CAP/LPM may also be appointed by the Director of OLES. The deputy P25 CAP/LPM shall be authorized to handle the work of the CAP LPM if recused or not available.

## 2.2 Laboratory Assessment Teams

The P25 CAP/LPM selects assessors on the basis of their professional and academic achievements, experience in the field of testing, management experience, training, technical knowledge, assessment skills, and communications skills. Assessors are assigned to conduct an on-site assessment of a particular laboratory on the basis of how well their experience, training, and skills match the type of testing to be assessed, as well as the absence of conflicts of interest.

**2.2.1** Team Functions

a) Conduct assessments on candidate and existing P25 CAP laboratories

b) Prepare reports based on assessments

c) Conduct monitoring visits on existing, recognized laboratories

d) Investigate and report on circumstances of any disputes

**2.2.2** Functions of Assessment Team Members

Assessment team members perform two related functions:

a)   The Lead Assessor (Laboratory Quality System Expert) applies detailed knowledge of the ISO/IEC 17025 standard and this handbook to evaluate a laboratory's management (quality) system.

b)   The Subject Matter Expert assesses the technical competence of a laboratory to the requirements of the technical standards and the requirements contained in this handbook.

**2.2.2.1**   *Requirements of Lead Assessors (Laboratory Quality System Experts)*

a)   A minimum of three days of formal training in the ISO/IEC 17025 laboratory accreditation assessment processes and one day in NIST Handbook 153.

b)   Experience having conducted ten or more prior assessments.

c)   Preference will be given to assessors with experience with a nationally- and internationally-recognized (that is, signatory to a Mutual Recognition Arrangement) accreditation body.

**2.2.2.2**   *Requirements of Subject Matter Experts*

a)   Familiarity with the test method document(s) including execution method for each test and the pass/fail criteria for each test.

b)   Familiarity with the necessary test equipment associated with each test.

c)   Familiarity with the normative reference document(s) (TIA TSBs, Compliance-Related TSBs, and Compliance Assessment Bulletins) behind the pass/fail criteria for each test.

d)   Familiarity with the parameters and methods for configuring and programming the device under test so as to generally discern whether the test operator has satisfactorily configured it.

e)   Familiarity with potential P25 feature interactions impacting the execution of a test or interpretation of the results of a test.

f)   A minimum of one day of formal training in ISO/IEC 17025 processes and one day in NIST Handbook 153.

**2.2.3**   Role of Lead Assessor

Each assessment team will have a Lead Assessor who acts as the single point of contact between the assessment team and the laboratory, and the assessment team and the P25 CAP/LPM. The Lead Assessor shall be, at a minimum, a Laboratory Quality System Expert per Section 2.2.2.1. The Lead Assessor may also be a Subject Matter Expert per Section 2.2.2.2.

**2.2.4**   The P25 CAP/LPM will provide the laboratory with a short biographical sketch of each assessor. If a conflict of interest or prior business relationship exists, a laboratory may request, and will be provided, an alternate assessor.

# 3. MANAGEMENT REQUIREMENTS OF RECOGNIZED LABORATORIES

## 3.1 Organization

**3.1.1** The laboratory shall carry out its testing activities in such a way as to meet the requirements of this handbook.

**3.1.2** The management system shall cover work carried out in the laboratory's permanent facilities, at sites away from its permanent facilities, or in associated temporary or mobile facilities.

**3.1.3** The laboratory shall:

a) Have managerial and technical personnel who, irrespective of other responsibilities, have the authority and resources needed to carry out their duties, including the implementation, maintenance and improvement of the management system, and to identify the occurrence of departures from the management system or from the procedures for performing tests and/or calibrations, and to initiate actions to prevent or minimize such departures (see also Section 5.2).

b) Have arrangements to ensure that its management and personnel are free from any undue internal and/or external commercial, financial and other pressures and influences that may adversely affect the quality of their work.

c) Have policies, procedures, and equipment to ensure the protection of the laboratory client's confidential information and proprietary rights, including procedures for protecting the electronic storage and transmission of results.

d) Have policies and procedures to avoid involvement in any activities that would diminish confidence in its competence, impartiality, judgment or operational integrity.

e) Define the organization and management structure of the laboratory, its place in any parent organization, and the relationships between quality management, technical operations and support services.

f) Specify the responsibility, authority and interrelationships of all personnel who manage, perform or verify work affecting the quality of the tests.

g) Provide adequate supervision of testing staff, including trainees, by persons familiar with laboratory methods and procedures, test methods, and the assessment of the test results.

h) Have technical management that has overall responsibility for the technical operations and the provision of the resources needed to ensure the required quality of laboratory operations.

i) Appoint a member of staff as Quality Manager (however named) who, irrespective of other duties and responsibilities, shall have defined responsibility and authority for ensuring that the management system related to quality is implemented and followed at all times; the Quality Manager shall have direct access to the highest level of management at which decisions are made on laboratory policy or resources.

j)   Appoint personnel to act in the absence of key managerial personnel (see Note below).

k)   Ensure that its personnel are aware of the relevance and importance of their activities and how they contribute to the achievement of the objectives of the management system.

NOTE: Individuals may have more than one function and it may be impractical to appoint acting personnel for every function.

**3.1.4**   Top management shall ensure that appropriate communication processes are established within the laboratory and that communication takes place regarding the effectiveness of the management system.

## 3.2   Record Keeping Requirements

**3.2.1**   Test Item Related Documents
The following documentation is required for each item tested at the laboratory:

a)   A detailed description of the item including hardware, software, and firmware.

b)   A copy of the test procedure(s) used.

c)   Identification of the test operator, the date of the test, and unambiguous identification of the unit tested (including the firmware version) and the infrastructure environment.

d)   A copy of the detailed test report including any applicable laboratory notes, intermediate computations, and analysis.

**3.2.2**   Laboratory Related Documents
The following documentation is required to substantiate the competence of the laboratory that conducted the tests.

**3.2.2.1**   *A descriptive biography for each laboratory employee detailing, for example, their years of experience, pertinent training or certifications, academic background, etc.*

**3.2.2.2**   *Equipment Records*
a)   Complete records on the latest calibrations for all test equipment in the laboratory.

b)   Configuration records for any radio infrastructure equipment used in the laboratory indicating the dates of any hardware, software, and/or firmware configuration changes on the equipment.

**3.2.3**   Retention Requirements
All documents related to a particular test object shall be held by the laboratory for as

long as the particular model of equipment is available for sale by the Equipment Supplier, including models whose software or firmware version numbers change over time, but at least for a period of five years. Records may be stored in hard copy or electronic format.

### 3.3   Laboratory Availability

Laboratories are advised to develop formal policies that are available to stakeholders describing the facilities' availability. Such policies might contain, for example, minimum requirements or expectations for invitational testing sessions, provisions for quick turnaround testing sessions, terms for minimum intervals between announced testing sessions, and periods of time for advance notice of testing. Any limitations to these policies should also be available for review by stakeholders.

### 3.4   Laboratory Testing Contract Requirements Review

The requirements that identify the system or units that will be tested should be carefully reviewed by Laboratory Technical Management. The test report(s) and the associated Supplier's Declaration of Compliance (SDoC), which the test report(s) support, shall apply only to the Model Class tested. Other configurations or variants that have not been tested shall not be included in the SDoC.

### 3.5   Subcontracting of Testing

Testing within the Scope of Recognition may not be subcontracted except to another P25-recognized compliance assessment laboratory with the same or complementary Scope of Recognition. Exceptions are the Federal Communications Commission (FCC) tests and other RF physical tests that may be subcontracted to an EMC test laboratory accredited by an ILAC signatory recognized by the FCC.

### 3.6   Purchasing Calibration Services

Purchasing of calibration services shall be controlled and specified by the laboratory to ensure that the order for services meets its specific requirements. The technical records from the calibration laboratory will be reviewed by the Laboratory Assessment Team to ensure they are traceable to NIST or the local National Metrology Institute. Reference NIST Handbook 150:2006 Annex B for further details on demonstrating traceability. As a minimum, the laboratory shall be traceable for the fundamental parameters of voltage, time, frequency, and RF power.

### 3.7   Access for the P25 CAP Laboratory Assessment Team

Upon request, the laboratory shall demonstrate to the P25 CAP Laboratory Assessment Team any of the procedures defined within the applicable Scope of Recognition for the laboratory.

### 3.8 Complaints

The laboratory shall record all complaints (from either internal or external sources) and process them using the corrective action and preventive action systems of the laboratory management system.

### 3.9 Control of Nonconforming Testing Work

If nonconforming work resulting in invalid test data are detected, the laboratory client and the P25 CAP/LPM shall be alerted and test methods repeated, as necessary, by the laboratory. If the nonconforming work is detected after either the test report and/or SDoC has been issued, the P25 CAP/LPM will investigate the nature of the nonconformity and will provide a written recommendation for corrective action to the laboratory and the laboratory's client. These parties shall have 30 days to respond to the proposed corrective action.

### 3.10 Improvement

The laboratory shall continually improve the effectiveness of its management system through the use of a quality policy, quality objectives, audit results, analysis of data, corrective and preventive actions and management review. Refer to ISO/IEC 17025:2005 Section 4.10 for further details.

### 3.11 Corrective Action

When departures from the laboratory's established policies and procedures are encountered, the laboratory shall have a defined process with clearly delineated responsibilities to implement corrective actions. Refer to ISO/IEC 17025:2005 Section 4.11 for further details.

### 3.12 Preventive Action

The laboratory shall have in place a system for identifying potential nonconformities and developing action plans to resolve such issues before their occurrence. Refer to ISO/IEC 17025:2005 Section 4.12 for further details.

### 3.13 Control of Records

The laboratory shall have in place documented processes for collecting, retaining, and accessing records related to the laboratory and the equipment tested. The laboratory shall retain test records for at least five years. Refer to ISO/IEC 17025:2005 Section 4.13 for further details.

### 3.14 Internal Audits

The laboratory shall periodically conduct internal audits of its own processes that shall include the management system, its past performance, compliance with this handbook, and laboratory processes that support P25 Compliance. Refer to ISO/IEC 17025:2005 Section 4.14 for further details.

### 3.15 Management Reviews

Laboratory management shall periodically conduct reviews of the management system. Refer to ISO/IEC 17025:2005 Section 4.15 for further details.

## 4. LABORATORY ASSESSMENT PROCESS

### 4.1 Application for Initial Recognition

**4.1.1** General

To initiate the laboratory recognition process, the applicant laboratory shall submit an application, a Quality Manual and relevant associated documentation specified within the application, and shall agree to conditions for recognition stipulated in the application.

**4.1.2** Required Information

An applicant laboratory shall complete an application for recognition that includes, but is not limited to, the following information:

a) The legal name and full address of the laboratory

b) The Authorized Representative's name and contact information

c) The names, titles and contact information for laboratory staff nominated to serve as Approved Signatories of test reports

d) An organizational chart defining relationships that are relevant to performing testing covered in the application

e) A general description of the laboratory, including its facilities and scope of operation

f) The requested Scope of Recognition as defined by the applicable relevant Compliance Assessment Bulletin(s)

Along with the application, the applicant shall provide a Quality Manual, applicable quality related documents, and other documents as may be required in future scope definition bulletins.

**4.1.3** By submitting the application, the laboratory's Authorized Representative commits the laboratory to fulfill the conditions for recognition listed in the P25 CAP Application Form. The Authorized Representative shall review all documents provided with the application package and become familiar with P25 CAP requirements before submitting the application.

**4.1.4**  Fees for Recognition
Reserved for future use.

**4.1.5**  Review of Application
Upon receipt of a laboratory's application for recognition, the P25 CAP/LPM assigns a Laboratory Code to the applicant laboratory; acknowledges receipt of the application in writing; reviews the information supplied by the laboratory for adequacy; requests further information, if necessary; and specifies the next step(s) in the recognition process.

## 4.2   Activities Prior to the On-Site Assessment

**4.2.1**  Assignment of Assessor(s)
A pool of assessors is selected by the P25 CAP/LPM based on their qualifications per sections 2.2.2.1 and 2.2.2.2. The P25 CAP/LPM then draws from this pool to create prospective Laboratory Assessment Teams. These teams are proposed to the laboratory's Authorized Representative for review per Section 2.2.4. The laboratory may request alternative assessor(s). The P25 CAP/LPM will recommend replacement assessors for the Laboratory Assessment Team to satisfy the Authorized Representative's concerns per Section 2.2.4.

**4.2.2**  Document Review

**4.2.2.1**  *The P25 Laboratory Assessment Team will review the laboratory's Quality Manual and related management system documentation submitted with the application to ensure they cover all aspects of the management system related to quality and, if followed, satisfy the requirements in this handbook. The Lead Assessor on behalf of the assessment team may ask for additional documentation pertaining, for example, to the management system, test processes or procedures, other supporting information (e.g., test configurations, list of test instruments used), and/or records in order to facilitate the review.*

**4.2.2.2**  *The P25 CAP Laboratory Assessment Team may identify suspected documentation nonconformities. These nonconformities are discussed with the Authorized Representative, and the laboratory is given the opportunity to address them prior to the on-site assessment. In some instances, based on the document review, the P25 CAP Laboratory Assessment Team may request that the laboratory address the nonconformities before the on-site assessment is scheduled. In such cases, the assessor will provide a list of the nonconformities to the laboratory in writing. Where the management system documentation requires significant revision, the P25 CAP Laboratory Assessment Team may require that the laboratory improve its documentation and submit it for further review prior to proceeding with the recognition process.*

**4.2.3**      Scheduling of On-Site Assessment

**4.2.3.1**    *The laboratory is contacted by the Lead Assessor of the P25 CAP Laboratory Assessment Team to schedule a mutually acceptable date for the on-site assessment. An assessment normally takes two to five days depending on the proposed Scope of Recognition. However, laboratory management should apprise the Lead Assessor of the laboratory's prior quality system experience and applicable qualifications, since this may affect the scheduled duration of the assessment. Every effort will be made to conduct an assessment with as little disruption as possible to the normal operations of the laboratory.*

**4.2.3.2**    *If a laboratory requires that its established assessment date be changed, it shall contact the Lead Assessor of the P25 CAP Laboratory Assessment Team.*

**4.2.3.3**    *An on-site assessment will be conducted as a part of the initial recognition process and every three years thereafter (based on the date of initial recognition). Delay of assessments beyond these frequencies may affect a laboratory's recognition status.*

**4.2.3.4**    *If a laboratory has a corrective action plan in place that includes corrective action to be taken within one year (as agreed upon by the P25 CAP/LPM) at the time of initial recognition, it shall have an additional on-site assessment within one year of initial recognition.*

Recognition may not be granted if any corrective action is not completed in an appropriate amount of time.

## 4.3    On-Site Assessment

**4.3.1**      Conduct of On-Site Assessment

**4.3.1.1**    *An on-site assessment may be conducted at all laboratory locations where P25 CAP tests will be performed.*

**4.3.1.2**    *At the beginning of the assessment, an opening meeting will be conducted with management and laboratory personnel to explain the purpose of the on-site assessment and to discuss the schedule for the assessment activities.*

**4.3.1.3**    *During the assessment, the Laboratory Assessment Team may examine equipment and facilities, observe demonstrations of testing, examine test reports, examine the management system, review quality and/or technical records and/or procedures, and review the biographies of staff to determine their competency in their particular area of expertise.*

**4.3.1.4**    *Laboratory Assessment Team members will use a common Test Method Review Summary, so each laboratory receives an assessment comparable to that received by others.*

## 4.4 The Post-Assessment Meeting

The post assessment meeting will provide a summary of all the activities of the assessment as detailed in Section 4.3. The nonconformities will be documented and presented to the laboratory. The laboratory will be given an opportunity to ask questions and clarify any of the nonconformities received. The laboratory shall have the right to object or seek clarification to any of the nonconformities. These objections will be handled by the P25 CAP/LPM.

### 4.4.1 On-Site Assessment Report

The assessment report will consist of evaluations of the quality system (in accordance with the applicable provisions of this Handbook) and of the technical assessment components. The technical assessment components will be contained in a Test Method Review Summary which details which tests were observed, what equipment was used, and the staff member(s) that conducted the test(s). In addition, the report will contain a listing of any observed nonconformities.

**4.4.1.1** *At the conclusion of the assessment, the Lead Assessor will conduct a closing meeting with the Authorized Representative (and other staff invited by laboratory management) to discuss observations and any nonconformities which are recorded in a written report.*

The report will include as a minimum:

a) The date(s) of assessment

b) The names of the assessor(s) responsible for the report

c) The names and addresses of all the laboratory sites assessed

d) The assessed Scope of Recognition

e) Comments and/or nonconformities cited by the assessor(s) on the compliance of the laboratory with the recognition requirements

f) A copy of completed Test Method Review Summaries

**4.4.1.2** *The Authorized Representative shall sign the report to acknowledge that the assessor has discussed its contents. See Section 4.5 regarding further steps in the nonconformity reconciliation process.*

**4.4.1.3** *The Lead Assessor will leave a copy of the report with the laboratory and forward the original report to the P25 CAP/LPM within five business days.*

**4.4.1.4** *The Lead Assessor is responsible for the content of the on-site assessment report, including the statement of any nonconformities.*

## 4.5 The Nonconformity Reconciliation Process

**4.5.1** Laboratory Response to the On-Site Assessment Report

If there are nonconformities listed in the on-site assessment report, the laboratory's Authorized Representative shall respond to the P25 CAP/LPM in writing within 30 days. In the event that nonconformities require more than 30 days to investigate, the Authorized Representative and P25 CAP/LPM will agree upon an appropriate response date.

**4.5.2** P25 CAP Laboratory Program Manager Response

The P25 CAP/LPM will respond to written communications from the Authorized Representative within 30 days.

**4.5.3** Ongoing Nonconformity Resolution

The Authorized Representative and P25 CAP/LPM will communicate with one another until a corrective action plan has been accepted by the P25 CAP/LPM or until all nonconformities are resolved to the satisfaction of the P25 CAP/LPM. Unless otherwise negotiated in advance, each party shall respond to the other's communications within 30 days or else the laboratory shall withdraw from the recognition process.

**4.5.4** Disputes

The Authorized Representative may appeal a finding or raise a dispute regarding this process in accordance with Section 4.1.5.

## 4.6 The Recognition Decision

**4.6.1** The P25 CAP may appoint a committee to assist in the decisions required for all actions related to granting, renewing, suspending, and revoking any P25 CAP recognition.

**4.6.2** The recognition decision process will consider the laboratory's record as a whole, including:

a) Information provided on the application

b) Results of management system documentation review

c) On-site assessment reports

d) Actions taken by the laboratory to correct nonconformities

e) If applicable, the adequacy of the corrective action plan(s) and preventive action plan(s) submitted by the laboratory

**4.6.3** All nonconformities shall be either resolved or adequately addressed in a corrective action plan to the P25 CAP/LPM's satisfaction before a recommendation for recognition will be submitted to the OIC CAP/PM.

NOTE 1: Only management system-related nonconformities (e.g., policies, systems, program, procedures, and instructions as described in Section 3 of this handbook will be accepted with only a corrective action plan as agreed by the P25 CAP/LPM. Technical (i.e., TIA-102) related nonconformities must be resolved to gain recognition. Laboratories recognized with a corrective action plan must provide quarterly corrective action reports which show significant milestones and continued progress in resolving identified nonconformities.

NOTE 2: Laboratories recognized with corrective action plans in place will be subject to an additional on-site assessment approximately 12 months after the initial recognition decision.

## 4.7    Granting Recognition

**4.7.1**    Recognition is granted when a laboratory has met all P25 CAP requirements. The renewal period is three years; recognition expires and must be renewed within three years of the recognition date (see Section 4.8).

**4.7.2**    Renewal dates which provide mutual benefit to both parties may be reassigned by the OIC CAP/PM or upon written request from the laboratory. If a renewal date is changed, the laboratory will be notified in writing of the change.

**4.7.3**    When recognition is granted, the OIC CAP/PM will provide the laboratory a Certificate and Scope of Recognition identified by its Laboratory Code, which includes:

a)    The name and address of the laboratory that has been recognized

b)    The Scope of the Recognition listing, the test methods for which the laboratory has demonstrated competence

c)    The laboratory's Authorized Representative

d)    The effective dates of the recognition

## 4.8    Renewal of Recognition

**4.8.1**    Each recognized laboratory will receive a renewal package containing an updated application form approximately six months before the expiration date of its recognition, to allow sufficient time to complete the renewal process.

**4.8.2**    The application for renewal shall be received by the P25 CAP/LPM prior to expiration of the laboratory's current recognition to avoid a lapse in recognition. If a laboratory allows its recognition to expire, the OIC CAP/PM may at his or her discretion require a new initial assessment.

**4.8.3**    On-site assessments of currently recognized laboratories will be performed in accordance with sections 4.3 through 4.7. If nonconformities are found during the assessment of a recognized laboratory, the laboratory shall submit a satisfactory response to the P25 CAP/LPM concerning resolution of nonconformities within 30 days of notification or face possible suspension of recognition.

NOTE: The on-site assessment associated with a renewal may occur at any time during the three year renewal period. However, it will normally occur within six months of receipt of the updated application. Renewed recognition is good for three years following review and acceptance of the application.

**4.8.4** Undue delay in the resolution of nonconformities may necessitate another on-site assessment.

## 4.9 Monitoring Visits

**4.9.1** In addition to regularly scheduled assessments, monitoring visits may be conducted by a P25 CAP/LPM authorized Laboratory Assessment Team at any time during the recognition period. The need for such visits will be determined by the P25 CAP/LPM. Monitoring visits will be scheduled in advance with the laboratory so that they do not interfere with scheduled or unscheduled interoperability testing events.

**4.9.2** The scope of a monitoring visit may range from checking a few designated items to a complete review. The Laboratory Assessment Team may review nonconformity resolutions, and/or verify reported changes in the laboratory's personnel, facilities, or operations.

**4.9.3** Laboratories are not assessed any fees for the cost of a monitoring visit.

## 4.10 Changes to Scope of Recognition

**4.10.1** A laboratory's Scope of Recognition may be expanded to include additional performance, conformance or interoperability Compliance Test Methods, including methods for additional P25 standard interfaces without the need for an immediate on-site assessment. If the laboratory requests additions to its Scope of Recognition, it must meet all the requirements defined in the applicable Compliance Assessment Bulletin for which recognition is sought. Laboratories may also make requests to reduce their Scope of Recognition.

**4.10.2** A laboratory wishing to change to its Scope of Recognition must make a request in writing. When requesting a reduction in its Scope of Recognition, a laboratory shall identify the relationship between the current and proposed Scopes of Recognition. When requesting an expansion of its Scope of Recognition, a recognized laboratory shall provide the following detailed information to the P25 CAP/LPM:

a)  Identify the relationship between the current and proposed Scopes of Recognition.

b)  Identify and evaluate the differences between the previously assessed Compliance Assessment Bulletins and the Compliance Assessment Bulletin(s) to be added under the proposed new Scope of Recognition.

c)  Summarize the critical parameters of the proposed new Compliance Assessment Bulletin(s).

d) Document special considerations contained in the proposed new Compliance Assessment Bulletin(s).

e) Identify any unique or new test equipment requirements.

f) Describe how staff competence with respect to the added Compliance Assessment Bulletin(s) has been achieved.

**4.10.3** This procedure is NOT intended to allow a laboratory to increase its Scope of Recognition to include Compliance Assessment Bulletin(s) for which performance, conformance or interoperability testing recognition has not been granted through an on-site assessment. The need for an additional on-site assessment will be determined by the P25 CAP/LPM on a case-by-case basis. A laboratory may also request deletions from its Scope of Recognition. The deletions may be temporary or permanent.

**4.10.4** Compliance with the requirements of the standards identified in the Compliance Assessment Bulletin(s) will be verified by a Laboratory Assessment Team at the next regularly scheduled on-site assessment of the recognized laboratory which occurs upon renewal of recognition.

**4.10.5** When a change to the Scope of Recognition is granted, the OIC CAP/PM will provide the laboratory a revised Certificate and Scope of Recognition (see Section 4.7.3).

## 4.11 Suspension of Recognition

**4.11.1** If it is determined that a recognized laboratory does not comply with the conditions of this Handbook established during the current assessment period, the laboratory will be notified in writing and given 30 days to respond. If the laboratory does not respond within 30 days, the OIC CAP/PM will suspend the laboratory's recognition. That determination may be made by the OIC CAP/PM (e.g., based on evidence obtained during the assessment process) or by the laboratory (e.g., by notifying the OIC CAP/PM of a major change in accordance with the application instructions). Suspension can be for all or part of a laboratory's recognition. If a laboratory's recognition is suspended for an extended period of time, the OIC CAP/PM may also propose to revoke recognition (see Section 4.12).

**4.11.2** If a laboratory's recognition is suspended, the OIC CAP/PM notifies the laboratory of that action, stating the reasons for and conditions of the suspension and specifying the action(s) the laboratory shall take to have its recognition reinstated. A reassessment of the laboratory may also be required for reinstatement.

**4.11.3** A laboratory whose recognition has been suspended shall not reference P25 CAP recognition on its test or calibration reports, correspondence, and advertising during the suspension period in the area(s) affected by the suspension. The P25 CAP will not require a suspended laboratory to return its Certificate and Scope of Recognition.

**4.11.4** A suspended laboratory may be reinstated by satisfactorily addressing the written concerns raised by the OIC CAP/PM. When recognition is reinstated, the OIC CAP/PM will authorize the laboratory to resume testing activities in the previously suspended area(s) as a recognized laboratory.

**4.11.5** During the suspension period, test reports generated by the laboratory shall not be used to substantiate a Supplier's Declaration of Compliance for any products.

## 4.12 Revocation of Recognition

**4.12.1** A laboratory that has had its recognition suspended for more than six months may be considered for revocation. If the OIC CAP/PM proposes to revoke recognition of a laboratory, he/she will inform the laboratory of the reasons for the proposed revocation and the procedure for appealing such a decision. Revocation can be for all or part of a laboratory's Scope of Recognition.

**4.12.2** The laboratory has 30 days from the date of receipt of the proposed revocation letter to appeal the decision. If the laboratory appeals the decision (see Section 4.15), the proposed revocation will be stayed pending the outcome of the appeal. The proposed revocation will become final through the issuance of a written decision to the laboratory, in the event that the laboratory does not appeal the proposed revocation within the 30-day period.

**4.12.3** If recognition is revoked, the laboratory may be given the option of voluntarily terminating the recognition (see Section 4.14).

**4.12.4** A laboratory whose recognition has been revoked shall return its Certificate and Scope of Recognition and shall cease to reference recognition by the P25 CAP in any of its reports, correspondence, or advertising related to the area(s) affected by the revocation.

**4.12.5** At the OIC CAP/PM's discretion, the laboratory's clients may be required to demonstrate that existing SDoCs are valid. Accordingly, the P25 CAP/LPM will work with the affected laboratory to identify whether any test cases must be re-run.

**4.12.6** If the revocation affects only some, but not all of the items listed on a laboratory's Scope of Recognition, the OIC CAP/PM will issue a revised Certificate and Scope of Recognition that excludes the revoked area(s) in order that the laboratory might continue operations in recognized areas.

## 4.13 Reinstatement of Recognition

**4.13.1** A laboratory whose recognition has been revoked may reapply (using the same procedures described in Section 4.1) and be recognized if the laboratory:

a) Completes the assessment process

b) Meets the P25 CAP conditions for recognition

## 4.14 Voluntary Termination of Recognition

**4.14.1** A laboratory may at any time terminate its participation and responsibilities as a recognized laboratory by advising the OIC CAP/PM in writing of its desire to do so.

**4.14.2** Upon receipt of a request for termination, the OIC CAP/PM will terminate the laboratory's recognition, notify the laboratory that its recognition has been terminated, instruct the laboratory to return its Certificate and Scope of Recognition, and to remove references to the P25 CAP from all subsequent test reports, correspondence, and advertising.

**4.14.3** A laboratory whose recognition has been voluntarily terminated may reapply per Section 4.1.

## 4.15 Appeals

**4.15.1** A laboratory has the right to appeal any adverse decision made by the P25 CAP/LPM or the OIC CAP/PM. Such decisions include refusal to accept an application; refusal to proceed with an assessment; corrective action requests; changes in Scope of Recognition; decision to suspend or revoke recognition; and any other action that impedes the attainment or sustenance of recognition.

**4.15.2** Appeals of decisions made by the P25 CAP/LPM are handled by the Director of NIST's Office of Law Enforcement Standards (OLES) or designee. Appeals of decisions made by the OLES Director are handled by the Director of NIST.

**4.15.2.1** *Appeals of decisions made by the OIC CAP/PM are handled by the Director of DHS' Office of Interoperability and Compatibility or designee. Appeals to decisions made by the OIC Director are handled by the DHS Director of the Command, Control, and Interoperability Division.*

**4.15.3** An advisory panel of experts selected by the P25 CAP/LPM may be called to address appeals of a technical nature.

**4.15.4** The party assigned to handle the appeal decides on the validity of the appeal and, if appropriate, renders a decision. The P25 CAP/LPM advises the appellant of the outcome of these deliberations and any recourse for further appeal.

## 5. TECHNICAL REQUIREMENTS FOR RECOGNITION

### 5.1 General

All of the test methods that are required within a particular Scope of Recognition shall be available at the laboratory.

## 5.2    Personnel

Staff shall be adequately trained for each test method or Test Case that they are responsible for or perform. Staff shall demonstrate proper programming, configuration, understanding and operation of the equipment under test and test equipment.

NOTE: Laboratory personnel may rely upon engineering support from the Equipment Supplier to demonstrate proper programming of equipment under test.

## 5.3    Accommodation and Environmental Conditions

The facility shall meet the minimum facility requirements specified in the test method standard. When specified in the standards, environmental conditions such as temperature, humidity, and barometric pressure shall be recorded at the time of the test. These records (measurements) shall be traceable to NIST or to the local National Metrology Institute.

## 5.4    Test Methods and Method Validation

It is the responsibility of each laboratory to validate each test method or Test Case in their own laboratory using the applicable equipment contained in that lab.

## 5.5    Equipment

5.5.1    Test equipment shall be available that can adequately perform the test case or test method.

5.5.2    All test equipment shall meet the requirements of the selected test method, including the normative standards referenced therein. This shall be demonstrated through product literature or actual measurements.

5.5.3    The laboratory shall have procedures for determining the proper operating condition of the test equipment.

5.5.4    All test equipment and equipment under test shall be configured in accordance with the selected test method standard unless an alternative method is employed, in which case, laboratory personnel shall provide the rationale for its use and explain how the results obtained using the alternative method compare with the method specified by the standard. If a test method has optional procedures, the laboratory shall indicate which option was used.

5.5.5    If modifications to the equipment under test setup are required to achieve a pass verdict, they shall be clearly indicated in the test report.

## 5.6    Handling of Test Items

All test items shall be suitably identified and stored in an appropriate place as required by the

Equipment Supplier. The laboratory shall have a procedure for determining the proper operating condition of the equipment under test.

### 5.7 Assuring the Quality of Test Results

To the extent possible, each test method and test case shall have a regular check to determine the validity of the indicated results.

### 5.8 Reporting the Results

**5.8.1** If available, the laboratory shall furnish one or more previously prepared detailed test reports or partial report samples prior to on-site assessment. The detailed test report shall provide:

a) The date and time that the tests were performed.

b) The location of the test facility (facilities that have received prior recognition from other conformity assessment bodies shall use the same address as was previously reported to the other body).

c) A list of all equipment tested.

d) Clear documentation of compliance with the applicable standard, including user information or labeling requirements.

e) The product identification and marketing name, installed software and firmware packages with revision numbers, version numbers, and serial numbers where applicable.

f) A list of the ancillary equipment and software required to configure the equipment under test, including revision numbers and serial numbers where applicable (this includes both equipment and software supplied by the Equipment Supplier and those created by the laboratory).

g) A complete list of test equipment used showing the arrangement of equipment and cables (drawings, photographs or block diagrams showing the interconnections between components are appropriate for this purpose).

h) Test equipment lists with manufacturer's model and serial numbers as well as date of last calibration and calibration interval.

i) All equipment set-up conditions and/or test equipment settings, so the tests could be repeated and give results within the tolerances defined for that measurement.

j) The measurement data in accordance with the standard in a clear and concise manner such as in tabular or graphical form (alternatively, chart data, instrument display captures or photographs may be used provided that the information is clearly presented and performance requirements are clearly indicated).

k) A summary which clearly indicates the test case verdict.

l)    The signature of the person(s) performing the test and the Authorized Representative.

**5.8.2**    All test reports shall follow the guidance given in the standard. If the test equipment for a particular test method is not configured in full conformance with the test setups described in the standard, the test report shall contain a complete description of the alternative arrangement. If an alternative test method procedure is employed, the test report shall provide the rationale for its use and explain how the results obtained using the alternative method compare with the method specified by the standard.

**5.8.3**    If a measurement procedure has multiple methods, the test report shall indicate which method was employed. Any and all deviations from test method procedures in the standards shall be recorded in the test report.